Learn
H

Contents

1) WHAT MAKES A HOME THEATER SYSTEM? 6

2) WHAT DO I NEED? 8

    THE DISPLAY SIDE OF THINGS.. 8
    SMART TV'S 9
    3D DISPLAYS 9
    THE AUDIO SIDE OF THINGS 10
    SOUND BARS 10
    SEPARATE SPEAKERS AND AMPS 11
    2 CHANNEL SYSTEMS 11
    MULTI CHANNEL SURROUND SOUND SYSTEMS 12
    SPEAKER SIZES 14
    SOURCE UNITS 15

3) THE BASICS OF SETTING IT ALL UP 17

    COMPONENTS THAT USE HDMI 17
    COMPONENTS THAT DON'T USE HDMI 18
    AUDIO AND VIDEO SOURCE UNITS 20
    WIRING UP SPEAKERS 21
    BI-WIRING YOUR SPEAKERS 24
    BI-AMPING YOUR SPEAKERS 25
    CONNECTING A SUBWOOFER 26
    SPEAKER PLACEMENT 30

4) CONNECTING OTHER DEVICES 32

    MEDIA PLAYERS 32
    COMPUTERS 33
    ANDROID SET TOP BOXES 35

- iPads and Android Devices 36
- What Is A 2ND Zone? 37
- Bluetooth 40

5) FINE TUNING 41

- Basics of Setting up an Av Receiver 41
- Other Receiver Settings 44
- Assigning Digital Inputs 44
- Network receivers 45
- HDMI Pass through 46
- Picture Scaling 47
- Input Naming 47

6) OTHER THINGS TO CONSIDER 48

- Universal Remote Control 48
- IR Kits 49
- Power Protection 50

7) TABLES TO HELP WITH SETUP 52

8) QUICK EXPLANATION GUIDE 54

- Receiver 54
- Source Units 54
- Surround Sound 54
- Digital Connection 55
- HDMI 55
- Sub / Sat Systems 56
- Floor Standing Speakers 56
- Bookshelf Speakers 56
- Center Channel 56
- Rear Surrounds 57
- Subwoofer 57
- Bluray 58
- Digital Co-Axial 58
- Digital optical 58
- Interconnects 58
- Universal remote 59

- Power Protection Boards 59
- IR 59
- RF 60

A COUPLE OF TIPS 60

- Is My Remote Control Working? 60
- Identifying A Speaker 61

9) END BLURB 62

Who Is this book for ?

This book is a basic guide to help those who want to move into the world of home theater, but aren't sure how to get there or how to put it all together.

It is to help people who want a non-technical idea on how to put their home theater system together and what parts are needed to do this.

There is no particular configuration you must have, or amount of money you need to spend to achieve this, everyone's system requirements and budgets are different.

Just because someone has a certain system and has spent a certain amount of money, doesn't mean you have to do the same to get the home theater system that you will be happy with.

This is not meant to be a technical guide to specs and features, it is a basic guide into what makes up a home theatre system for people that are wanting to step into this area and upgrade from say just having a television in their living room with just a dvd player plugged into it.

I do not work for any of the brands I mention in this book, but will refer to some brands throughout this simply because I have experience with them. This is not to say they are the best for your needs, they will be used as a reference simply because I have experience with them and their features and setups.

Dedicated to Tina-Marie and her patience with our ever changing

Home theater system and the ongoing new products she has to master.

1) What Makes a Home Theater System?

Home theater system is a broad term used to describe audio visual products connected together in your own home.
Originally it was to bring you closer to the experience of going to the movie theater.
You get to enjoy better sound and picture quality and an overall movie experience.
Now days though there is so much more to a 'Home Theater' system as new products like Tablets, Games Machines, Video Calling and so much more become the norm and we are wanting to integrate them into our entertainment systems.

These days your home theater system also can double as a great music system and even allow you to easily connect devices such as cameras, media players, computers and games machines to mention but a few to widen the experience.

Show off your holiday movies and pictures on your big screen Television when friends come around without all having to huddle around the computer screen.

Home theater systems can be anything from a television and a couple of speakers to a fully fledged multi channel, high definition picture and sound system with a projector, or one of the many options in between.

In general, there is no right and wrong combination of products or setup as everyone has different requirements and budgets.

One thing to establish early on is the balance of movie watching to music listening that you require from your system, as this will influence the products that you will require in order to establish this goal.

Its far simpler to create a system that gives you a satisfying movie experience compared to a system that sounds great for listening to your favorite music, this is because its relatively easy to make sound effects whiz around the different speakers in your room whilst a lot of your attention is on the screen, there is a lot going on to fill your senses.

When you are sitting down and listening to music only, you tend to be more critical as there is no video or images so you are solely focused on the music quality.

In my opinion and what I always say to a customer, if music quality is important to you and will be what your system will be used for the most, Audition speakers and amplifiers in a music setting, by this I mean sit down and listen to the speakers that interest you, and play some of your own music that you are familiar with.
Once you are happy with how the combination of amplifiers and speakers sounds playing music there is a far greater chance that they will also sound fantastic in surround sound watching a movie, but be sure to audition your system doing both.

2) What Do I Need?

The Display Side of Things

The majority of people who are wanting to move from simply having a television in their house with maybe a dvd player plugged into it, to a home theater type setup in a living room will opt for a large flat panel display as these are a very important part of the visual experience and can be purchased for a reasonable price these days.
Size of this depends on your room size and of course your budget anything from 32" to 65" or bigger! In Plasma or LED / LCD displays are usually the option for the visual side of your setup.
I won't get into the 'Which is better' argument as this is not what this is about, that is for you to audition and decide what you are happy with.

(Of course you can opt for a projector as well if you have the room and jump up to a 100" screen or more as well if room size and budget allow!)

The display is the final item in the chain of components and a very important part to the overall performance of your setup, so choose a panel that you have actually sat down and watched, as everyone's perception on what is a great picture / performance is different!

Don't base your purchase on reviews of what others say - It is a good idea to read <u>some</u> reviews on a product, but don't get too caught up in reviews, as in my experience if you look hard enough you will find a bad review on pretty much everything out there, as someone for one reason or another will have had some negative experience.

This doesn't just apply to televisions either; apply this to pretty much every product you buy!
With a new flat panel display, the key is to actually 'watch it' and 'use it' and see if you are happy with its performance, ease of use and quality.

Smart TV's

Something else to note if you are getting a new flat panel display, these days there is a lot more to a television than just playing broadcast television programs.
New 'smart' televisions can also connect to the internet and let you do things such as use Skype to talk to friends via video chat and also browse the internet, plug in USB hardrives to allow recording of programs broadcast, watch on demand services and playback of photos / videos and much more, So make sure you talk to the salesperson about these other options if they interest you.

It is easier to have these features integrated into the television than try and add them via another device later on.
Keep in mind though that the more of these 'Smart' features that you use, you will also consume more data, so make sure your broadband plan has enough data to allow for these extra features to use. These services will require that your display is connected to the internet ether via Ethernet cable or via Wifi.
(This is model dependant though and some require optional accessories to connect them to your internet connection)

3D Displays

3D is another option available in modern flat panel displays and also in some projectors, things to consider when looking at 3D as an option are:
You need to have a source unit that can play 3D, a 3D Bluray player would be the most common example of this and are available for a very reasonable price these days.

You need to have the fore mentioned display 3D capable and you also (if using one) will require your Av receiver to be able to pass 3D content through it and up to the display, You will need compatible 3D glasses as well at this stage for viewing and lastly the Hdmi cables connecting all these components together need to be able to pass 3D over them as well. You will require all of the above to be able to experience 3D on your home theatre system.
(HDMI is currently required for 3D; you cannot experience 3D on your display using analogue connections.)

The Audio Side of Things

Once you have decided on your new display you can move onto the sound side of your system.
This can be simply a couple of speakers and an amplifier, through to a 5.1 or 7.1 surround sound system.

Sound Bars

If you simply want to improve the sound of your flat panel television for watching television and dvd or Bluray discs, (as sometimes the speakers in these can let them down) and don't want to have to have speakers everywhere around your living room making it look like an audio showroom, Then you could consider a 'Sound bar' or sound projector as they are referred to.
Many brands now produce these and they consist of a slim line speaker system similar in length to your new flat panel display and connect to it.
This allows for vastly improved sound quality through better and bigger speakers and their own built in amplifier, than is found in most flat panel displays these days.
Also some makes / models have a separate subwoofer (some are wireless as well!) which will vastly improve the bottom end or bass of your system adding depth and

realism to your system without the need for separate amps and large speakers.

There are lots of options even within the field of 'Sound Bars' so look carefully into features that some of these have above simply making the tv sound better.
Most sound bars are an easy setup and are physically placed under the flat panel display itself and connected to the display via an Optical Cable and or HDMI cable.
It's an easy 'Out the box' solution and will suit quite a lot of people who want to upgrade the audio of their new flat panel television for a reasonable price.

Separate Speakers and Amps

The next step up from a sound bar type upgrade is to go to a separate amplifier and speakers,
You can either opt for:

2 channel systems

This is where you have an amplifier, which has the ability to connect your DVD, television, games console etc and is connected to a pair of speakers.
This is a big step up in audio performance and will instantly make your gaming or movies sound so much better than just the television speakers!

There are many different speakers and 2 channel amplifiers available, but even a basic 2 channel amplifier connected to a pair of floor standing type speakers will transform you watching and listening experience greatly!

This will not let you experience a true surround sound but is a great improvement to your home theater setup if you currently only use the speakers in your television.

A 2 channel system as this is called will also double as a great system for listening to music.
This is where you need to decide if your setup is going to be just for watching movies or listening to music as well. Generally speaking it is a lot easier to make a system sound great for watching movies than it is to make it sound great for listening to music.

Multi Channel Surround Sound Systems

This is the next step up and will require the addition of an amplifier commonly known as a 'receiver' or 'home theater Av receiver'
These have a lot more going on under the hood than a standard 2 channel amplifier; they become the hub or brains of your home theater system.

All your devices such as Bluray, Dvd, Set top boxes, Games machines etc connect to this receiver and then it connects to your display.
They do all the amplification of the audio and send it to your speakers as required; they decode the audio tracks on a DVD, Bluray or other multi channel sound source and send it to the appropriate speakers, so that you hear the effects and voice etc as it was meant to be.

They also have many other features depending on make and model you choose, they can do things such as allow you to control them via a Smartphone or tablet App, Play a separate source to another room at the same time,
Upscale a video signal to a higher definition to improve its quality plus much more.
This of course varies with the brands and the models with in the brands, so check on what the model you are considering has to offer if you want more than just your basic Av receiver.

We will talk about this a little more further on.

A multi channel surround system can be made up of many different configurations,
The most common is a 5.1 Channel system.
This comprises of
2x front speakers, left and right which are your main speakers and do the majority of the work
1x Center channel speaker, that is responsible for most of the voice or dialogue in a movie,
2x rear effects speakers that help to create the effects whizzing by you, when say a helicopter flies over your head in a movie and finally a Subwoofer that is responsible for the bass or low effects used in movies to add depth and realism to a soundtrack.

The subwoofer is the .1 part of a 5.1 soundtrack, 5.1 means there are actually 6 speakers, 5 of them are powered off the amplifier and the .1 or 6th speaker is traditionally the subwoofer, these mostly what is known as 'Active' meaning that they have their own amplifier built into them as the traditionally large speaker needed to reproduce the bass requires a more powerful amplifier than could be built into a home theater receiver.
(There are of course other options to this but we are going to focus on active subwoofers as they make up the majority of subs in home theater systems)

However, you can settle for less speakers if you don't want to have so many speakers around your room, you can opt for just two fronts (2.0) , 2 fronts and a centre (3.0), 2 fronts and a sub (2.1) or two fronts, a centre channel and a sub (3.1) or the full 5.1 or even 7.1, the latter adds 2 more effects speakers known as surround backs.

With a modern Av receiver you simply need to tell it in the speaker setup menu what configuration you have and it will

then do all the required processing to make the most of your particular setup.
This makes it nice and simple if you are building your system up and adding additional speakers as your budget allows, for example if you start with a 2.0 system and then buy a subwoofer, all you have to do is go back into your speaker setup menu and tell the receiver that you now have a 2.1 system instead of 2.0 and it recalculates this for you sending the low effect sounds to the subwoofer now instead of your front speakers.

Speaker Sizes

Even within speakers you can choose from many options to suit your requirements, anything from small satellite systems, known as sub sat packs to large floor standers or even complete in ceiling or in wall speaker systems. (Even in wall subwoofers are available but they are more of a custom install product so we won't talk about them here, we will stick to more traditional free standing subwoofers.

In my opinion if you can fit them in your budget and they are acceptable in your room, Go for a floor standing type speaker, especially if music listening is going to be a big part of your new system as these are the easiest way to create a nice full sound.

Sub / Sat packs where you get 5 small speakers for fronts, rears and center and also a matching subwoofer are a great way to go if you want a stealth clean look to your room as these are available in different colors as well with some brands so they can be wall mounted and pretty much disappear once installed, these are great for watching movies but may let you down a little If you really like to listen to your music as a 2 channel system.
However even within the sub/sat systems there are a few brands that do a pretty good job being a music listening

system as well as a surround sound movie system so if you are considering this type of system be sure to listen to it playing a movie in surround sound and also playing 2 channel audio from a CD etc.

Picture of a typical Sub / Sat (or Satellite) system

In Ceiling speakers are another option if you want a truly minimalist look, providing you can get cables to all your speakers this is another great option to consider, once again I would tend to recommend them more for a movie experience but a set of flush mounts for fronts and rears and one as center channel can make for a nice discreet surround sound system.

You would still need to add a subwoofer to this ideally as well to complete a 5 or 7.1 setup.

Source Units

Source units are the devices you play or send content from, such as DVD players, Bluray players, CD players, set top boxes, turntables, games machines such as Xbox, PlayStation, Media players and even computers and tablets such as iPads etc.

All these devices are connected to the Av receiver which in turn decodes the audio and video and send them to the speakers and display as required.

Connecting your devices this way also means that as you change inputs on the Av receiver from say DVD to Xbox, it will switch the video from that device up to the television,

meaning you no longer have to remember what Av channel the television has to be on to watch a certain source.
(This is based on your receiver being of a modern variety that does video switching, pretty much all receivers now a day's do this)

More and more devices now are connected by a cable known as 'HDMI', this is the latest type of connection allowing both Audio and Video to travel from one device to another over one single cable, HDMI is also the standard connection for allowing transfer of High Definition video and audio content.
This alone makes for connecting your source equipment such as a Bluray disc player to your Av receiver very easy! As all you need to do is have one cable from the HDMI out of the Bluray player that goes into one of the HDMI inputs on your Av receiver and that's it!
All audio and video pass through this one connection.
(In case you're wondering HDMI stands for High Definition Multimedia Interface)

If you have an older source unit that you still want to use but it doesn't have an HDMI output, you can usually still integrate it into your system, but will have to use alternative analogue outputs from that device and connect these separately to your Av receivers Analogue inputs to allow it to process audio and video signals. (This is model specific as some Av receivers can do more with signal conversion than others)
Signal conversion and scaling are available on some Av receivers and they can use processing circuits to enhance a standard analogue video signal and 'Scale' it up to a higher definition – the result is a better quality image on your HD display from a lower quality content.
For example a movie you filmed on a older video camera can be 'Up Scaled' to a better quality picture for viewing on your new HD display

3) The Basics Of Setting It All Up

Now that you have all your shiny new products sitting there, it's time to get them all hooked up and talking to each other.
We are going to base this mainly on modern day current products as this of course is what most people are currently buying.

Components That Use HDMI

Most sources units these days have an HDMI out terminal (we will cover some other connection types further on) if your sources unit (these being Set Top boxes, Games Machines, Bluray and Dvd Players etc) do have HDMI Out then the process is fairly simple.

HDMI as we stated earlier can pass both audio and video over it, so connecting your source unit to your new Av receiver is fairly easy and straight forward.

Take for example that you have a new Bluray player, on the back of this source unit it will have an HDMI socket, this will be the output that the audio and or video comes out of the Bluray player on, Simply use an HDMI cable of preferred length and plug one end into the Bluray players HDMI out and then connect the other end of this HDMI cable into one of the HDMI inputs on the back of your receiver, make a note of which one you have used as commonly these are just labeled HDMI 1, HDMI 2 etc on a receiver, as you will need to know later on which HDMI input you have used so as to switch the receiver to the right one to watch your Bluray discs.
(Some Av receivers such as Yamaha have an HDMI input labeled Bluray for example, if yours has this then choose that HDMI input for your Bluray)

That is all that is required for the AV connection side of most source units that has HDMI out!

To watch this source unit - In this case your Bluray player, you turn your Av receiver to the HDMI input that you plugged this cable into and the receiver decodes the signals for audio and video and plays them to the speakers and screen.

Apply this same method for all source units that use HDMI, remembering along the way which ones are allocated to each source unit.

Once all the HDMI sources are connected, you will need to use another HDMI cable that will go from the Av receivers HDMI output up to the Screen and into one of its HDMI inputs. (HDMI cables are not directional)

Once this is done, you only need to select on your display that particular HDMI input you have chosen and whenever you switch the Av receiver from one source to another the receiver decodes and displays the video and plays the audio from that particular source, No need to remember what Av channel the display has to be on any more as anything connected to the receiver uses this same HDMI in on the display!

Components That Don't Use HDMI

If you have other source units that don't have HDMI outputs, this could be because they are audio only devices such as CD players or they may be older products before the days of HDMI such as an older DVD player, These can still be used with most modern day receivers.

Audio Only Source Units

If for example it's an audio only device like a CD player, then you can connect them via Digital or Analogue connections, The simplest being the latter, this only involves a pair of left and right cables known as interconnects, these normally will use RCA or Phono plug type connections and are a stereo configuration meaning

there is one for the left channel and one for the right channel.
Using a pair of interconnects the correct length you plug left and right into the CD players left and right outputs (normally colour coded as white (for left) and red (for right) and then on the other end of the interconnect goes into a pair of Analogue inputs on your Av receiver.
On the Av receiver look for a pair of the same looking terminals as the CD player had and labeled as an input, these are not always labeled as CD but look for this in case they are, if there are none labeled CD (in) they may simply be labeled 'Input 1' Input 2' etc.
Once again make a note of which input you have chosen for this connection as you need to switch the Av receiver to this particular input when wanting to play this source, (In this case the CD player)

If you want to use a digital connection from your audio source (we will for this take it that it is a CD player, but it could be any audio source with a digital audio output) There are a couple of different digital audio connections.

1) Digital Co-Axial - single cable that has RCA Phono type connections on both ends
2) Digital Optical - Single Cables that has specialized connections that only fit other Optical digital cables / sockets
Either is perfectly fine to use and in my opinion neither is better than the other, However you may want to at this point check on the back of your Av receiver to see if it allows for both Digital Optical and Digital Co-Axial inputs.
Once you have established this and got the correct cable, plug one end into the sources digital audio out (CD player in this instance) and then the other end into one of the Digital inputs on the Av receiver, once again make note of what this digital input is assigned to or labeled on the back of the receiver.

We would use a digital connection in for example the following scenario, Say you have an older CD player - maybe 15 years old and a brand new Av receiver, chances are that the DACS (Digital Analogue Converters) on the 15 year old CD player won't be as good as the brand new Av receiver, so we use digital out on the old model CD player and use the DACS on board the receiver to decode the digital stream of audio on the compact disc you are playing, this basically bypasses everything in the Cd player except the mech spinning the disc and the laser reading the digital audio and feeds it to the superior decoders on the Av receiver.
End result = CD player sounds as good as it possibly can as it is now using more modern Digital to Analogue converters and preamp circuits.

Audio and Video Source Units

If the source unit plays audio and video but has no HDMI output, such as say an older model DVD player, you should still be able to connect this to your new AV receiver but you need to use a combination of Analogue video connections and digital audio connections.(for surround sound)
Firstly there various analogue video connections, there is the S-Video connections (similar looking to a keyboard plug on a PC), A component Video connection (has 1x Blue, 1x Red and 1x Green separate RCA connections) and then the very basic Composite video connection which is simply just a single RCA type connection usually terminated Yellow in colour.
In terms of the best quality - I would rate them in the following order;
Lowest - Composite Video
Middle - S-Video
Highest - Component Video

Have a look on the back of the source unit and see which one of these yours has available, then check on the back of your Av receiver to make sure that it allows for the same inputs.
S-Video is being dropped from a lot of Av receivers now so you may not find these on yours, so look for component or composite inputs, these will normally be labeled as AV 1 In, AV2 in etc.

Once again make a note of which input you are using for this particular source as you will need to know this later to view this source unit on your display.
For the audio side of things if your source unit has a digital out terminal as we described above for the CD player then use one of those, as this is the only option to get a digital audio signal from this source to your Av receiver.
For example if it's a DVD player and you want to play it in Dolby Digital 5.1 then you must use either Digital optical or Digital Co-Axial as these are the only other connections that can pass a multi channel audio stream to your receiver.

If you only have 2 speakers connected to your receiver or amplifier then you can simply connect the Analogue 2 channel left and right outputs from the source to the amplifier (same way we described above for the CD player)

Wiring Up Speakers

When it comes to wiring up your speakers it is very important to make sure that you carefully get all the negative terminals and positive terminals connected correctly, this is known as keeping everything in 'Phase'.

To make sure everything is in phase, try to use speaker cable that has one conductor marked, this is usually done with one of the sides of the cable having a stripe or a '+' symbol on it or it might be a different colour or shape completely.

The most common would be to have for example, a standard (2 core, 1 Negative & 1 Positive) speaker cable grey in colour, one side is grey and the other is grey with a black stripe running along it, Take one of these as the positive and one as the negative and keep this the same the whole way through your speaker configuration for both the speaker end and the amplifier end of your connections.

For example I always take the stripe on a cable such as mentioned above as the negative (Black terminal on a speaker or amplifier usually), As long as you keep this system the same for all speaker connections your speakers will be all running in phase correctly.

If you do get speakers running out of phase it means that when a pair of speakers is playing the same audio, one speaker would be pushing and the other pulling at the same time, the most obvious result is that the low frequency or Bass gets cancelled out making music sound hollow or prominent in the midrange.

As a tip for a pair of front speakers for example, listen to some music and turn the balance all the way to one side and listen to the bass notes of the music, if it sounds like there is more bass than there is when you turn the balance back to the middle then you probably have one of these speakers out of phase.

Some Av receivers these days are able to tell you if a speaker is out of phase when you use its Auto setup system, this enables you to easily get your speakers wired up correctly and get the best out of your system.

There are a few different ways you can connect your cables to your speakers and the majority of them these days have what is known as binding posts on the rear of them, these allow you to simply screw the terminal down on top of the copper of your cable or preferably use a Banana plug to connect.

Your speakers may use one of the following options

1) Spring loaded terminals – Simply strip your cable back, open the spring clip and insert the copper under the clip
2) Binding posts – No banana plug, simply undo the screw terminal and insert the stripped cable under the screw terminal and tighten it back down.
3) Binding posts – with banana plug, have your cables fitted with banana plugs then push them into the end socket of the binding post (this is the preferred method if you have this as it provides better connection and a bigger surface area of contact.)

Some speakers also have more than one pair of terminals on the back of them as shown in the picture below of a Castle Knight floor standing speaker, this allows for some optional ways to connect such as Bi-wiring and Bi-amping.

If you wish to just connect with a standard 2 core cable from you amplifier connect it as shown in the picture below, this picture shows a speaker that allows for both standard and Bi-Wire / Bi-amp options but is connected with a standard 2 core speaker cable (with QED banana plugs in this instance)

If you wish to use Bi-Wire or Bi-Amp cabling then remember to take out the bridging links which are the bent metal strips connecting the 2 sets of negative and positive terminals together.

Bi-Wiring Your Speakers

If you want to Bi-Wire your speakers, providing they allow for this with their connections, you need 4 core cable that has 2 positive conductors and 2 negative conductors, they are then commoned or 'joined' together at the amplifier end

(Both negatives are joined together and both positives are joined together at the amplifier end.)

This type of connection means you are separating the inputs to the crossover in the speaker and using dedicated runs of cable for the high frequencies and the low frequencies.

Bi-Amping Your Speakers

If you want to Bi-Amp your speakers, providing they allow for this with their connections, you need 4 core cable that has 2 positive conductors and 2 negative conductors, they are run as separate cables and the amplifier must be able to be switched to Bi-Amp mode, this is in effect now using 4 amplifier channels to drive this pair of speakers, one for each low frequency input and one for each high frequency input of the speaker pair.

You must also remove the linking bars for this mode as well; if both your amplifier and speakers have the ability to run in Bi-Amp mode this is the optimum way to power your speakers.

Why Bi Wire or Bi-Amp ?

The theory behind these two options to wiring up your speakers is quite different,

Bi-Wiring means that you still have a common connection at the amplifier end of your cables and this splits out to a pair of cables for the high frequency of your speakers and another pair for the low frequencies of your speakers, to do this you remove the link bars as shown in the picture above, the theory behind this is that the low and high frequencies are kept apart from each other resulting in cleaner sound as one does not disrupt the other as then run from amp to speaker.

Bi-Amping means that there is two complete separate cables running from the amplifier to each speaker, one for the low frequency part of the speaker and another for the high frequency part of the speaker, this allows for the same result as listed above for Bi-Wiring and also means that you are in effect running one amplifier channel for each low and high input for each speaker.

This results in better control and more power to your speaker as each amplifier channel is only running one part of your speaker.

(You need to establish that your amplifier can be set to Bi-Amp mode 1st though as this requires some configuration normally before you wire them up in this manner)

Connecting a Subwoofer

Connecting a subwoofer to a modern Av receiver will in most cases require just a single cable that has a Phono (RCA) plug on each end.

This is a single cable as the subwoofer channel is a mono signal; it requires a low level preamp signal which comes from the Subwoofer Output on the back of the Av receiver.

Ensure that you use a proper 'Subwoofer Cable' as using a cable that doesn't have proper shielding can result in unwanted noise being induced in to the subwoofer – Normally mains hum which results in loud low frequency hums coming from your subwoofer.

In the picture to the left of a typical active subwoofer for a home theater system you will see that there are left and right inputs, there are 3 things you can do here

1) Use a single subwoofer cable and use the Left RCA input terminal
2) Use a female to Male RCA Y Adaptor and plug that into both left and right inputs terminals.
3) If you are connecting to an amplifier that has no subwoofer output but has a pair of left and right preamp outputs, you can run a stereo pair of subwoofer cables to both the left and right RCA inputs on the subwoofer and use the built in crossover on the subwoofer (See below)

The output terminals would be used for connection to a 2nd Subwoofer for example if you want to run 2 subwoofers and only have one Subwoofer preamp output on your Av receiver.

You will also notice that on this particular subwoofer there is a pair of speaker binding posts, these are for connection in a system where the amplifier has no preamp / or subwoofer preamp outputs.

To use these you would run a pair of speaker cables from your amplifier to these speaker level inputs and another pair to your speakers from the same terminals on your amplifier you would in this case also use the built in crossover on the subwoofer.

Crossovers on your Subwoofer: You will see that on the back of this subwoofer there are two dials.

One is the crossover for the subwoofer and the other is the Level (basically this is another volume control)

If you are using an Av receiver then you would be using the crossover built into that device so you don't need to set the crossover point on the subwoofer. (Leave it set up as high as it goes as the receiver sets the crossover in this case)

The crossover in an active subwoofer tells it what frequencies to reproduce, For example if you set your crossover dial to 100 Hertz, the subwoofer will filter out any audio above 100 hertz and not pass it to the sub to be played, and all audio from 100 Hertz down will be played through the subwoofer.

The point of this crossover is to match where your main speakers stop producing audio effectively or what's called 'roll off' and tell the subwoofer to come in at this same point.

Example, your front speakers are cut off at 100 Hertz so set the subwoofer to around the same 100 hertz so that between the front speakers and the subwoofer you are getting the whole frequency range but letting the subwoofer do the very low audio of which it does a better job.

A good starting point with subwoofers like this, if you're using a pair of floor standing speakers would be to set the crossover at 100Hz (either on the Av receiver or on the sub depending on your setup) and the level on the back of the subwoofer at about half way and try your system.

Listen to some music and or a movie you know well and see if it sounds right, you can always vary the volume and or crossover until it suits.

One thing to note when setting up a subwoofer is always make sure that it is sitting in its final place in the room when you try it as they can be easily influenced by their placement in the room and distance from walls etc.

At this point you can also try the 'Phase' switch or dial and try it in its different settings as this also can correct a subwoofers sound depending on where it is and where it's facing in the room.

If there is also a switch on your subwoofer that says something along the lines of 'Auto / On /Off' normally you would select 'Auto' as this detects the audio when it arrives and tells the subwoofer to turn on, this is a power saving feature and also a nice convenience as you don't have to get up and physically turn he subwoofer on or off.

Very important thing to mention here is always plug the low level cable in at both the subwoofer end and the amplifier end BEFORE you connect the mains and turn it on.

Speaker Placement

Where your speakers are physically placed in a room can influence their sound and therefore consideration as to where they can fit and look right must be taken into account at the beginning when you are looking at speakers to purchase.

Think about things like, how can I get the wiring to a certain speaker? Can I run it under the house? Can I get cabling into a wall or through a ceiling without having to renovate the room? Does the cable have to run past a door way? These things must all be taken into consideration.

Every speaker will need a cable run to it and in the case of a subwoofer it will need a low level subwoofer cable and also mains power (normally) as well.

There are some wireless speaker solutions on the market but keep in mind that even a so called wireless pair of speakers will need power to make them work, if you need to get power to them, you may as well just run speaker cable!

Most of us don't have a dedicated Home Theater room, so we make do with our main living area or lounge; Because of this most rooms aren't of the perfect size and layout.

Thankfully the modern Av receiver used in a surround sound home theater system can adjust and correct for this to a fairly large extent.

In its most basic setup a 5.1 surround sound setup will consist of 2 front or main speakers that do the majority of the work (when playing in surround sound mode) and all of the work in a 2 channel system, these are your front left and front right speakers, try and get a decent spread (distance) between these 2 speakers to allow for good stereo imaging. (At least a couple of meters between them in my opinion)

The center channel does a fair bit of the work in a surround sound system and is responsible for the dialogue or vocals when watching a movie in 5.1 or 7.1, this speaker needs to be centrally located with the television or screen that you are using and as close to it as possible.

For example you have a 42" flat panel television sitting on a cabinet, the center speaker ideally needs to be located under the TV and also in the center of the TV, this is difficult in some scenarios so try to get it as close to this position as you can. The receiver can correct for this to a certain extent which we will cover later on in the 'Setup area of this book'

Rear surround speakers, these are placed behind you and also consist of a left and right channel in a 5.1 setup, these are used for effects and would do the least work in a typical surround sound setup, however they are still important as they are used or example to create the effect of objects moving past. You have a bit more flexibility with these speakers and they can be stand mounted, wall mounted or even a pair of in ceiling flush mounted speakers, once again the receiver can correct to a certain extent for their placement but make sure that they are at a minimum above the height of your couch or seats that you are sitting in to watch your movie.

Subwoofers, these can be influenced by their placement in the room and also their particular configuration, for example a rear ported speaker will change in its bass response depending on how close to a wall it is placed, A subwoofer can also be affected by this and also by placing it in a corner.

If a subwoofer sounds to boomy, try moving it away from a wall or corner and try it in different places until you get a satisfactory result, Also with subwoofers being a low frequency speaker they aren't so critical to their placement in relation to your ear, low frequency is not as directional as the high frequencies coming from your other speakers so you have more options as to where you would like your subwoofer in the room.

4) Connecting Other Devices

It is of course possible to connect other devices and this is becoming more common these days.

People want to integrate their Pc and other multimedia source units to their home theater systems as well.

We will briefly talk about some of these below here:

Media Players

Media players are a type of set top box that allow playback of many multimedia file formats, A common example of a Media player would be a device that allows connection to your Av receiver via an HDMI cable and also has built in USB terminals for connection of USB hardrives, USB keydrives and SD cards etc

You plug it into one of your HDMI terminals on your Av receiver, Switch the receiver to that particular input and whatever is played from the Media Player box comes up on your display and plays through your speakers.

These are very handy and easy way to integrate playing movies or pictures from your camera and also for playing content such as video files from a hardrive.

Pretty much all these media players come with a remote to control them and you navigate through the menus via the display (normally your television)

Some also have Wifi or Ethernet connections that allow you to connect to networks or the internet as well to play files from other devices on the same network.

Some also have internal hardrives allowing you to store content for playback all on the one device.

If you are archiving video files (for example, home movies shot on your HD movie camera) on a hardrive attached to your media player, keep in mind that you will need a lot of storage to save video files, so try and invest in the biggest hardrive or USB key drive that you can otherwise you may soon find you are running out of storage space. This really is a case of biggest is best when looking to purchase!

Computers

Connecting your Pc or any computer to your home theater system makes for an awesome experience!

Imagine easily being able to bring up your browser on your television screen and look around the internet all from the comfort of your couch!

Connecting your computer to your home theater system can bring about some side issues as well, so be prepared to have to try different settings on things like your video card in your computer as sitting a few feet away from the television screen brings about problems such as font size and being able to read your browser results etc as opposed to sitting only a couple of feet away from the monitor on your desk.

Once again this is getting easier with modern computers also featuring HDMI terminals, If you have this option on the computer that you wish to integrate with your home theater system then this is by far the best way to go as the Av receiver then treats the computer the same as if it was DVD or Bluray type of source unit, One HDMI cable does all the video and audio and a lot of new televisions are able to resize the desktop so that it fills the screen properly.

This is by far the easiest way to connect a computer to your Av receiver.

If you don't have HDMI you can still do it but I strongly suggest either, getting a video card with HDMI out or at least DVI out and using a DVI to HDMI converter.

You can use a standard VGA monitor type connection in some instances to connect your computer to your television (this is if your television or display has the required VGA input terminal) and then use the audio output from your computer to your Av receiver but this doesn't allow for all the newer higher definition settings.

You can always talk to your local computer store and enquire about upgrading to a video card that allows for HDMI out if you want to make it as simple as possible.

Also things to consider, you will need some type of preferably wireless keyboard and mouse, Some brands now have keyboards with built in trackpads on one side similar to what is on laptops and that gives you one device that is wireless that you can control your computer from the lounge suite with.

Personally I ended up going down the road of using an Apple Mac Mini, these are well suited to being integrated into a home theater system due to their size and specs such as HDMI and Digital audio out etc (please note not all Mac Minis have HDMI, some of the older models require an adaptor) Team this with a Bluetooth keyboard and touchpad and you have an awesome computer experience on your big screen television.

Android Set Top Boxes

Another option that can give you a similar experience to having a computer connected to your system is an Android powered Set top box,

These are much the same as a Media player set top box to connect but are powered by the Android operating system so it can give you many more options such as browsing the internet and using video chat programs, accessing your email and much more.

This is a great solution particularly if you're on a budget but want to have some of the features you would get with connecting your computer to your home theater system.

Once again these days a lot of these have built in Wifi and USB for connecting external drives and also feature HDMI outputs meaning an easy integration to your system.

Don't forget with this option you will also need a keyboard and mouse device – preferably wireless to control it all !

iPads and Android Devices

If you have devices such as an Apple iPad and or iPhone you can also integrate these into your new home theater system, however you will normally require some additional hardware for some features.

Many of the latest Av receivers on the market allow you to use an Apple feature called 'Airplay' this allows you to push content from your Apple device (iPhones, iPads and some iPods) wirelessly to the receiver.

To do this you're Av receiver needs to have a network feature and be connected to the same network as your iPad or iPhone is.

Network receivers allow you to connect them to your home network via either Ethernet or Wifi (some require a separate device for wireless)

So, you plug your Ethernet cable into your Av receiver, this is in turn connected to the router in your home, you then connect your iPad or iPhone etc to the same network over Wifi and that's it, you can open up your music on your device and use the 'Airplay' option to wirelessly push your music to your Av receiver.

Very cool!

While we are briefly talking about devices such as iPads and iPhones there is another feature available on most new Network capable receivers that allow you to control the Av receiver via an App.

You can do this on both iOS Apple devices and also on Android devices such as android tablets etc but you would need to check 1st if your brand of Av receiver has

downloadable Apps for your particular device.
The benefit of using this is that you have a far better remote control for you receiver, as the App is easier to read and laid out better than most remote controls are. It can also operate your receiver or other source unit from another room and unlike most infrared remotes doesn't require 'Line of Sight' to work.

Imagine for example that you have a 2nd zone (another room that has speakers in it running off the same receiver), you can walk in to that room and control the volume and inputs wirelessly. This is especially useful when your 2nd zone is a pair of outdoor speakers on a deck for example.

Simply take your device with you (if it's your phone it's probably already in your pocket) and once out on the deck you can use it to control the volume without having to run back inside to turn it up or down.

What Is A 2nd Zone?

A 2nd Zone or a 'Powered Zone' is a feature that some higher end Av receivers can also allow you to music in another zone.

Think of this like another complete audio system in another room or 'Zone', the benefits of this is that still using the same Av receiver that you use in your main room for watching movies and or listening to music you can also supply audio (and in some cases video as well) to the other zone.

For Example, Family are watching a movie in the lounge in surround sound, You have a 2nd Zone which is a pair of speakers out on your deck, You can be sitting out there enjoying the sunshine and listening to music from a source such as the Tuner, iPod, Cd player etc.

You have your own volume which is completely independent from the family watching the movie and your own control over what source you are listening to, and if you have a network receiver and a control App you can easily use this to turn your Zone on and off and change your music and volume.

You now have two complete music systems all operating from the one Av receiver.

If your receiver has the ability to do this, you must set it up initially in the Setup menu, To do this bring up the onscreen menu on your Av receiver and find the section to do with 'Zones' or amplifier configuration, you will need to look for something along the lines of setting your receiver to have an 'Powered 2nd Zone'

The most common Av receivers that are used for this are 7.1 channel receivers, they have a special mode that you can effectively take 2 channels that would be otherwise dedicated to 6th and 7th channels for surround backs on a 7.1 system and assign them to a 2nd Zone

Depending on your particular Brand and model of Av receiver you may only be able to run 5.1 and a 2nd zone, others will allow you to still run a 7.1 and a 2nd zone but will switch off the surround backs and dedicate them to the 2nd zone if the scenario occurs where a 7.1 soundtrack is being played in the main zone and someone comes along and switches the 2nd zone on all at the same time.

You will need to check on the documentation of your particular receiver to see how it handles these scenarios.

Also one thing to note in regards to 2nd zones is that most Av receivers will only allow you to send analogue signals to the 2nd zone, by this I mean signals from sources such as a Tuner, Cd player, Turntable iPod (depending on its specific connection) etc, This is because the receiver would need another set of DACs (Digital to Audio Convertors) on board to be able to decode digital signals and send them to the 2nd zone. (This is not the case for all receivers and is model / brand specific)

So if you are wanting to listen to a source such as a Bluray player connected to your Av receiver via HDMI and can't seem to get the audio to come out at your 2nd zone, this may well be the case, If this is indeed what the problem is, in the case of a source unit such as a Bluray / DVD player, you can also look at connecting a pair of analogue left and right cables from the analogue out of the player to an analogue input on the back of your receiver and select that particular input for your 2nd zone.

Analogue and digital signals normally appear on source units like a Bluray player simultaneously, so this can be a good way around this potential problem.

Also note that there are some Av receivers out there that have 3rd and even 4th zones all on board, potentially giving you more multi room audio options all from one amplifier unit.

Some will also let you send video as well as audio to another zone, meaning a cheap way to have 2 complete audio video systems in your home all from one Av receiver.

You will need to carefully read what your particular brand and model of receiver is capable of doing if other zones is something you want to use. Talk to your sales person if considering purchasing a new receiver to make sure you get the exact features you need.

Another simpler option that is available on receivers normally slightly below models that offer a true 2nd zone can be what is known as 'B Speaker Switching', this will allow you to have another pair of speakers in the same room / another room and turn them on and off separately to the main speakers, However they can only play the same source and have the same volume as the main speakers.

This may suit you though and generally will be a cheaper option in the range of receivers compared to a 'Powered 2nd Zone' model.

Check that your particular model can do both A and B speakers at the same time, as there are some around that can do either A or B but not both at the same time.

Bluetooth

Bluetooth is another wireless connection system that is incorporated into home theatre and audio systems these days.

With a Bluetooth enabled receiver or speakers you can wirelessly send music from devices such as portable music players, Phones and laptop computers.

To use this system, you need to 'Pair' your music source for example a phone with Bluetooth to the receiver or Bluetooth enabled speakers, Turn the Bluetooth on in the settings of your phone and search for Bluetooth devices, it should then find the receiver or device that you are wanting to send music to, select to pair to this device and follow any instructions to accept the connection (sometimes a 4 digit code is required for this, so refer to your owner's manual or the pairing code that appears on your phone)

Once the connection is established you can play music from your device and stream it via Bluetooth.

Even if you have an amplifier that has no Bluetooth option, You can buy a separate Bluetooth module that accepts the signal from your phone, laptop etc and converts this back into analogue so that you can connect to a standard audio left and right Input and presto you have a Bluetooth enabled amplifier !

5) Fine tuning

Now comes the tuning up part of your system to make sure you get the best from your new Av products.

With a new system if you have purchased a new Av receiver, you will need to carefully go through setting up some of the basics 1st before you sit down and listen to music or watch a move.

Turn your receiver on and find which button you use on the remote to go into the 'Setup' usually with modern day receivers they can show the setup menu on your television or display that you are using, this is known as OSD (On Screen Display), if not it will appear on the front display of the receiver.

Basics of Setting up an Av Receiver
1) Telling the receiver what your speaker setup consists of, This is very important to do 1st before you try playing music through your new system, If your Av receiver has the option for an automatic setup and came with a setup Microphone, plug this

in and follow the instructions in your owner's manual to activate the auto setup option.

What this will do, is go through and play test tones to determine and setup your system for speaker size, speaker configuration (5.1, 7.1 or 2.1 etc), individual channel level and also the time alignment process.

Normally in this case you plug your setup microphone in and place it in the position where you will be sitting to watch movies, this then sets up all the above parameters for optimal listening in a surround sound environment.

If you don't have an auto setup option as above or you want to do it manually, then in the 'Setup menu' you need to do the following.

2) Select and tell the receiver what speakers you have in your system, for example if you have large floor standing type speakers for your front speakers, you will select 'Front Large', If you also have a center channel, rears and subwoofer then you go through these options one by one and tell the receiver what size these particular speakers are, If you don't have all of these speakers, you select 'None' for that particular speaker selection.

This tells the receiver what makes up your system and how to process multi channel audio tracks and where to send them for the best sound reproduction.

This is most important to do especially if you are using small bookshelf or satellite speakers, before you trying playing audio through them, otherwise you may damage a small speaker by sending low frequencies to it that should be going to a subwoofer for example.

3) You must also go into the amplifier configuration and tell the receiver if you are using things such as a 2nd Zone or Bi-Amp if these are options on your model and you are using these, It must be done before you try playing audio through your system.

4) Time Alignment, If you are setting up your Av receiver manually, you must also setup the time alignment, this tells the receiver the distance you are sitting from all the particular speakers that make up your system.

So take a tape measure and carefully measure the distance from where you sit to watch a movie to each speaker one by one,
Measure from your seating position to the front left speaker and write this down, then from your seating position to the center channel and write this down, do this for all speakers that you have in your system.
Then find in the setup menu for speakers in your Av receiver where you can enter these distances in, once you have carefully gone through all speakers in your system, you have now done your time alignment settings.
This is important as it determines how the receiver sends the audio from the various speakers to arrive at your ears at the correct time, this makes for

effects etc to sound how they were meant to when recorded.

5) Channel Level Adjustment, When you are manually setting up your Av receiver you may also need to adjust the individual channel levels, Most Av receivers will allow for you to adjust the level or volume of each channel independently, this is useful for example if one speaker such as a center channel is less efficient than the front speakers and you are finding it difficult to hear what people are saying when talking in a movie.

You can go into the speaker level adjustment part of the Av receiver setup and go through to the particular speaker that you want to adjust and adjust its level up or down, this can be done for fronts, rears, surround backs, Centers and subwoofers.

This is a bit of a personal thing as well as some people like louder effects, more bass from the subwoofer or louder dialogue from the center channel etc, so feel free to play around with this to suit your personal taste.

Other Receiver Settings

There are many other settings that you're particular Av receiver may have available so we will try and cove some of them here.

Assigning Digital Inputs

Sometimes the need arises where you need to assign a digital Co-Axial or Digital Optical input to a particular Av input,

For example – You have a Media player connected to your Av receiver via Component analogue video connectors, this takes care of the video side of things when watching your Media player, but you want to be able to also use the digital output of the media player for the sound side of things. Component video input on your receiver maybe AV4 for example, you then need to tell the receiver that you want to 'Assign' or "steer" that particular Digital Audio input to AV4 so that when you select AV4 on your receiver the Video comes through from the Media player and also the digital audio comes through.

This is normally found under the 'Input' section of the receiver setup

Network receivers

If your Av receiver is a Network enabled receiver you can use it to stream content from another device on the same network, For example you may have music stored on a computer that is connected to the same network, you can play music located on the computer through your receiver which is in another room, Or another example that is found on some receivers is Internet radio which allows you to connect to internet based radio stations and stream the music they are playing via the internet to your receiver.

Another common application for connecting your Av receiver to your home network these days is to take advantage of controlling it via an App, We discussed this in section 4 but if you're Av receiver has an App that you can download to a device such as a tablet or phone then you can use this to control the Receiver, Particularly handy if you have a 2nd zone and are out of sight of the receiver and want to change the volume etc.

To do this your device is normally connected via Wifi to the same network that the Av Receiver is also connected to.

In most cases it is simply that we have a wireless modem / Router in our home that we connect to the internet with, This same router is connected to the Av receiver via hardwired Ethernet cable or via Wifi and the controlling device being a tablet or phone for example is connected to the same network via Wifi, this enables all the devices to see each other on the network.

HDMI Pass through

You may come across a setting called HDMI Pass through in your receiver settings as well, this is a setting that can allow the following for example.

Say you watch TV through a set top box like Sky for example, this is connected via HDMI to your receiver and the receiver to your television also via HDMI, If you want to just watch the news and not have to turn on the receiver to do this, with HDMI pass through enabled, the sound and video will bypass through the receiver and go straight to the television and the sound will play through the built in speakers on the television.

This can also be useful for the kids watching a cartoon DVD for example on the television, you don't want to have them playing with the Av receiver so HDMI pass through allows the dvd audio and video to go straight through the receiver and play on the television.

Picture Scaling

Another setting you may see in the HDMI section of a modern Av Receiver is 'Scaling' this is to do with what your receiver can do in the way of manipulating a video signal.

This can be 'Up Scaling' where it can take an Analogue (and in some cases digital) video signal and uses it's on board processors to scale this image up to a higher resolution resulting in a better image quality.

Other settings under this area may include a particular setting for the scaling such as, Upscale to 720P or upscale to 1080P, 'Through' which means the receiver just passes the Video signal through without doing anything to it, Or 'Auto' Which will select what setting its finds best for the signal.

The wording on this will vary from brand to brand slightly but they are fairly straight forward to work out, most are set on Auto by default and won't need to be changed

Input Naming

Input naming is simply the option to rename a particular input to something that makes more sense to you, For example you may need to switch through Av inputs AV1, AV2, AV3, AV4 etc to watch the different sources that you have connected to your receiver.

Having these renamed to things such as 'Sky Tv', Bluray Player' Play Station' etc make a lot more sense in a real world when other members of the family are using the system and although it's a simple thing, If your Av receiver can do this, I recommend doing it.

6) Other Things To Consider

Universal Remote Control

Other things to consider when you have your new system up and running could be the addition of a universal remote control, Once you have a few different components making up your home theater system you will also end up with a good sized collection of remote controls, This can be confusing for everyone and something that I recommend is to get yourself a 'Learning Remote' or Universal remote as they are also known as.

I use Logitech Harmony remotes, and although I have no association with them, I find them excellent to setup and use and do recommend them.

What a universal remote allows is for you to have one remote that learns the controls of all the devices that make up your system, this frees you up to only have the one remote out on the table and all the original remotes can stay in the draw.

With the Logitech Harmony series of remotes for example, they are programmed on your computer (Pc or Mac) and require an internet connection and for you to set up a free account with Logitech, once that is done, you simply tell the software that comes with the remote, what make and model of products you have and it adds them to your setup.

Then using a system called Macros or 'Activities' you can setup commands such as 'Watch TV' and the remote tells the TV to turn on and turn to a certain Av channel, any other device needed such as a set top box or a Sky TV box to turn on and if required your Av Receiver to turn on and turn to a certain input, Easy as that, then the volume up and down on the Universal remote becomes the volume for watching TV

It makes for using your new multi component home theater system as easy to use as selecting an activity like 'Listen to Cd' Or "Watch DVD' and the remote does all the tricky stuff for you.

IR Kits

Ir kits or Infrared kits are something to consider when you want to hide your electronics away from sight in a cupboard or cabinet or even in another room; these allow you to still control all of your devices via the Infrared remote controls whilst not having the devices in line of site.

These consist of a 'IR Receiver', A Hub and the IR emitters, Some brands sell a complete kit that has all of the parts required and depending on how many devices are to be controlled determines which kit you would need so make sure the one that you are looking at has enough IR Emitters for all your devices.

You position the IR receiver in line of sight so that any remote control commands sent to it are received and sent to the IR hub which is hidden away out of sight, which in turn sends the IR signals out to the emitters which then respond to the various commands.

There are many different options when it comes to IR receivers to suit your particular requirements, you can purchase table top mounted ones, Flush mounted ones or even ones that are wall mounted in a plate.

Another use for IR control like this is for use in a 2nd Zone; this will mean that you can still have full control over your system from the other zone even though there is no line of sight for a remote to traditionally work.

Say for example that you have a pair of speakers out on a deck and this is your 2nd Zone, Once you are out on the deck you can't control the system which is in another room inside the house. With an IR system installed you can simply take the remote or a 2nd remote with you and point it at the IR receiver mounted outside in the deck area and that will send all commands back to the system inside, allowing you full control.

IR systems generally require the running of Cat5 or Cat6 cable (sometimes called Data cable) to the area where you want to mount the IR receiver.

Power Protection

Another device that you should allow for in your budget is a power protection board

This device allows you plug the mains supplies for all your devices such as your Av receiver, DVD player etc into it and that in turn has one plug that goes to the mains socket in the room where you are setting up your home theater system.

This device protects against power spikes, brown outs, power cuts and other fluctuations in the mains power supply that can potentially damage your sensitive electronics, some of these power boards do more such as conditioning the power supply (smoothing the mains) as well, so have a look at what the different brands and models offer in the way of protection.

Some models also offer protection for your phone lines and aerial connections as well, by allowing you to route the phone and aerial connections through the protection board as well as the mains supplies for your separate devices.

Others also have the individual power sockets switched and labeled which makes it easy to isolate one particular device if you want to disconnect it from the mains power.

7) Tables To Help With Setup

Write down in the table below, each of your speakers and the distance to them, this can be used as a reference if you setup your time alignment manually in your Av receiver

Speaker Distance Measurement Table

Speaker	Distance
Right Front	_____
Left Front	_____
Centre	_____
Left Surround	_____
Right Surround	_____
Surround Back L	_____
Surround Back R	_____
Sub	_____

This is the measurement from each of the speakers listed above, to the listening position where you will be sitting to watch your movies.

This is provided to make an easy reference for you if you wish to setup your time alignment manually in your Av receiver speaker settings.

Av Input Table

Use this table to note which source units are connected to which Av inputs on the back of your Av receiver

Source Unit	Input on Receiver
(DVD, Sky TV Etc)	(Av1, HDMI 2 Etc)
_____	_____
_____	_____
_____	_____
_____	_____
_____	_____
_____	_____
_____	_____
_____	_____
_____	_____

This is provided for you simply so that later when using your system, it is easy to look up which input on your Av receiver is needed to watch a certain

Source, For example, you want to watch a Dvd, to do this you may need to switch your Av receiver to HDMI 2 and then to watch another source such as a set top box for cable TV you may need to switch your receiver to AV 1

This is also handy if you want to program a universal remote or learning remote to operate your system as you will need to know what inputs all your source units are connected to.

8) Quick Explanation Guide

Receiver

A receiver is the name given to an amplifier that has a tuner built into it, and in the case of a Home Theater receiver this also can process all the audio and video signals coming from your source unit.

Think of this as the brains or Hub of your home theater system, everything connects to it and it does all the work and sends the audio out to your speakers and the video out to your display.

There is also 2 channel receivers that are an amplifier and tuner in one unit but only designed for 2 channel audio mainly and don't normally have the ability to switch and process video signals (Although there are also exceptions to this)

Source Units

Source units are simply the device that plays the media (or files) that you wish to listen to or watch.

These are typically things like DVD players, Bluray disc players, Set top boxes, games machines, media players etc, When thinking of connections typically think of the signal coming 'Out' of a source unit and 'Into' a receiver or amplifier, this helps to avoid getting inputs and outputs confused when wiring up your system.

Surround Sound

Surround sound is achieved by having a certain amount of speakers placed around you and the amplifier sending audio to them at precise times so that the audio appears at your ear as was intended when it was recorded.

This can come in many different arrangements, 5.1 being the most common that I deal with, 5.1 refers to 6 channels of audio, typically but not only found on DVD and Bluray soundtracks, The '5' tracks in a 5.1 recording are as follows: 2 for front left and right speakers, 2 for rear effects left and right speakers, one for center channel dialogue speaker and the 6th or '.1' track being the low effects or subwoofer channel. The subwoofer in the most typical home theater setup is an 'Active' speaker meaning it has its own amplifier built into it; this is done as the typically large sized speaker needed to reproduce bass notes needs more power than would be able to be available straight out of an Av receiver.

Digital Connection

A digital connection is required by a home theater receiver to decode the surround sound, sound track, this can be done over HDMI, Co-Axial Digital or Optical Digital, It is not possible to send a digital signal from a source unit to an Av receiver via analogue cables, so to experience true surround sound you must use one of the above digital connections.

HDMI

HDMI or High Definition Multi Media Interface is the only current connection that can carry high definition audio and video over the one cable, this making for a relatively easy setup when connecting a source unit to an Av receiver.

This connection can carry standard definition and high definition, so is perfect for DVD players and Bluray players; HD set top boxes and Gaming machines to name a few.

Sub / Sat Systems

Sub / Sat systems, are typically a 5.1 speaker system consisting of 5 small 'Satellite' speakers for fronts, rears and center channel and a matching subwoofer that has been made to compliment the natural high roll off of the small satellite speakers, these are great for space saving as the very small speakers take up very little space, These systems are great for small rooms and movie watching. You normally would buy a system like this all in one carton as a matched set and add an Av receiver to complete your setup.

Floor Standing Speakers

Floor standing speakers are the larger of home audio speakers, that are free standing and generally good for music listening and movies.

If your system is going to be used more for music or mainly for music and you can fit in a pair of floor standing speakers then in my opinion this is what I would recommend.

Bookshelf Speakers

Can also be used for music listening, generally the smaller brother of the floor stander they are suited for fronts and rears and can be mounted on stands or cabinets in your room, Some allow for mounting brackets so that you can wall mount them as well. Often used as a rear speaker in a 5.1 system and most brands offer them as a matching smaller brother to their floor standing option.

Center Channel

The center channel speaker is the speaker situated as close to the display (normally a television) and used mainly for the dialogue or voices in a multi track surround sound system such as a 5.1 DVD, the purpose of this is to create the effect of the voice you hear coming from the person on the screen talking.

Rear Surrounds

Rear surround speakers are the effects speakers in a 5.1 type setup and allow for the transition of sound through the speakers making a sound effect such as a helicopter flying over your head possible by the effect starting at say the front speakers and gradually fading through to the rear speakers, because every speaker in this setup has its own channel of audio it is possible to create the effect of an object going past you.

Subwoofer

A subwoofer is a speaker dedicated to reproducing the low frequencies or bass effects of a soundtrack; this is either from music or movie soundtracks. In the case of a multi channel digital audio track such as a 5.1 movie, the .1 part of the soundtrack is a low effects or bass track.

Subwoofers are typically larger speakers and in the most common variety are what we call 'Active' subs, this means that the subwoofer speaker comes with its own amplifier and is all housed in the same cabinet, this is because typically more power is needed to push the larger subwoofer speaker than would be available from an Av receiver,(there are variations to this but an active sub is the most common used in a home theater system)

A subwoofer is a very important part of a 5.1 type system and should be carefully considered when purchasing your setup.

Bluray

Bluray is a type of optical disc that has become the current standard of media for High Definition video and audio, Bluray discs have a far greater capacity than standard DVD discs and use this to hold the huge amounts of data needed for high definition video and audio recordings.

You cannot play a Bluray disc on a standard DVD player; However a Bluray player is backwards compatible and can indeed play standard DVD discs as well as Bluray Discs. The Bluray name is derived from the 'Blue' laser needed to read these discs, Normal DVD and CD discs use a 'Red' laser. (DVD actually stands for Digital Versatile Disc, Not Digital Video Disc as if often thought)

Digital Co-Axial

Digital Co-Axial is one form of digital connection that uses Phono type RCA plugs on each end and is capable of transferring multi channel digital sound tracks from a source unit such as a DVD player to an Av receiver.

Digital optical

Digital Optical is another form of digital connection that uses its own dedicated connectors on each end and is capable of transferring multi channel digital sound tracks from a source unit such as a DVD player to an Av receiver. This system uses light pulses and is encoded at the source end and decoded back again at the Av receiver end.

Interconnects

Interconnects is the name given mainly to a pair of analogue left and right stereo cables that would typically be used between a source such as a CD player and the amplifier, it is important to use a decent level of cable between these two components so as not to make a sub standard cable the weak link in your audio system.

Universal remote

A universal remote or 'Learning Remote' as they are also known as, is a remote control that can learn all the commands from all your different devices making up your home theater system, this means that you only have to have one remote control out on the table instead of multiple ones.

Power Protection Boards

Power protection boards are a multi-box power supply that allows you to connect all the mains power plugs of your home theater devices such as your DVD player, Av receiver etc to it and it in turn is plugged into the mains socket in your home theater room, They offer protection to your devices from potential electrical damage caused by power spikes, brown out and black outs. (Well worth the investment for your sensitive electronics)

IR

IR is the term used to describe infrared signals, this is what is used by most remote controls to send their commands from the remote to the device, For example when you push the button for volume up on your Av receiver remote control, a signal is sent via IR to the Av receiver which decodes the signal and turns the volume up, IR is line of sight so the signal needs to be visible to the device receiving the command.

RF

RF is another term sometimes associated with remote controls, this is far less common than the IR system mentioned above and uses Radio Frequency to send a command from the remote control to the receiving device, there are some aftermarket remotes that use this technology and they come with a module to convert the RF signals back to IR signals so that your existing devices can be controlled.

You would use this as an option for when the devices that you wish to control are not in line of sight, IE you have your Av receiver etc hidden away in a cupboard or in another room.

A Couple of Tips

Is My Remote Control Working?

Not sure if your remote control is working at all ?, Could just be a flat battery or a faulty remote but it's hard to tell if a normal old IR remote is working or not as we can't see infrared signals with the human eye,

However if you have a web camera on your pc turn it on and look at the little IR LED that is in the end of the remote through live camera image on your screen then press a button on the remote control, You will see the IR LED flash a white colour if it is indeed emitting IR signals.

Hold down a button on the remote and you should see a series of white flashes, if not then the remote is not working, Check batteries 1st !

Identifying A Speaker

Once you have all your speakers wired up and run back to your amplifier, it's often hard if not impossible to tell which wire is coming from which speaker.

An easy way to identify which speaker is on the end of a particular cable is to take a 1.5Volt AA battery and momentarily touch one of the cables across the battery terminals (negative of the cable on one end and positive on the other)

(Don't leave it on there, just quickly touch it a couple of times across the terminals), you will hear a 'Click' come from the particular speaker that is attached to this cable, Get someone to help you by listening to which speaker the click comes from, This makes getting the correct speaker connected to the correct output on your Av receiver much easier!

9) End Blurb

The images used in this book were used with permission of Capisco Ltd

All brand names and trademarks mentioned in this book remain the sole property of their respective owners.

♦

I have worked in the electronics industry for more than 25 years and specifically in the audio industry for more 18 years.

Having Trade Certificate in electronics servicing I later moved into the sales area of home (and car) audio systems in which I still currently work.

The ever changing world of consumer electronics has meant that there is always something new to learn and keeps me passionate about the industry.

This book has come about because of the commonly asked questions that I have encountered in my day to day dealings with customers looking to setup their own home theatre systems.

© 2013 Allan Turner

Made in the USA
Middletown, DE
10 July 2017